Water
220

鲸的菜园
The Whales' Garden

Gunter Pauli

[比] 冈特·鲍利 著

[哥伦] 凯瑟琳娜·巴赫 绘

田 烁译

上海遠東出版社

特别感谢以下热心人士对童书工作的支持：

匡志强　方　芳　宋小华　解　东　厉　云　李　婧
刘　丹　熊彩虹　罗淑怡　旷　婉　杨　荣　刘学振
何圣霖　王必斗　潘林平　熊志强　廖清州　谭燕宁
王　征　白　纯　张林霞　寿颖慧　罗　佳　傅　俊
胡海朋　白永喆　韦小宏　李　杰　欧　亮

目录

Contents

一只磷虾正环顾四周，她看到了大量的浮游生物。比起几天前，食物多了许多。磷虾喊道：

“鲸回来啦！鲸回来啦！”

“噢，这样的话，食物链的循环又开始了。”一只浮游生物叹息道。

“循环？这不只是循环，这可是世界级的饕餮盛宴！”

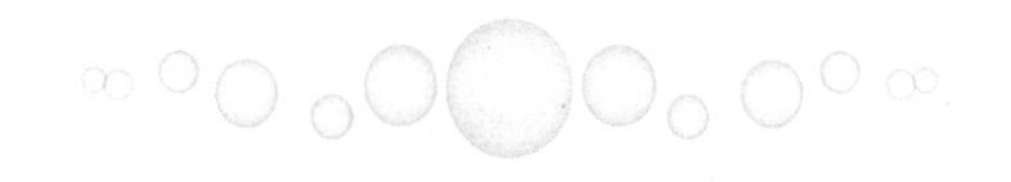

A krill is looking around and sees an abundant amount of plankton. There is so much more food than just a few days ago. The krill shouts:

“The whales are back! The whales are back!”

“Oh, so the cycle starts again,” one plankton sighs.

“Cycle? This is not just a cycle, this is the greatest food feast in the whole wide world!”

鲸回来啦！

The whales are back!

大的吃小的……

The big ones eat the smaller ones ...

“盛宴？你管这个叫盛宴？大的吃小的，小的吃更小的？”

“这根本不是大吃小的问题。这是关于生命的问题——为了所有海洋生物的利益，我们如何通过保持物种的多样性来丰富海洋的生命。”

“那么，就说说鲸是怎么丰富你的生命的吧？”浮游生物问道。

“A feast? You call this a feast? When the big ones eat the smaller ones, that in turn eat even smaller ones?”

“This is not at all about size. This is about life – and how we enrich the ocean with ever more life for all.”

“And just how are the whales enriching your life?” Plankton asks.

“每次他们大便，我们就会得到大量营养，一切就都有了生机活力，一切。”

“我知道。当他们巨大的粪便被排入水中时，它会给每个生物带来强大的动力，尤其是像你我这样的微型物种。”

“是的。粪便富含铁元素——那正是我们生长繁衍所需要的。”磷虾回答说。

“Every time they poop, we receive a shower of nutrition, and everything comes to life. Everything.”

“I see. When their mighty poops are released in the water, it gives everyone a massive boost, especially the tiny ones like you and me.”

“Yip. That poo is so rich in iron – exactly what we all need to grow in size and in numbers,” Krill replies.

粪便富含铁元素……

That poo is so rich in iron ...

他们可以促进浮游生物大量繁殖……

They create blooms of plankton ...

“那么，你是不是在告诉我，因为鲸吃掉了我们，所以他们可以促进浮游生物大量繁殖，进而使得你们磷虾越来越多地出现在我们身边，然后鲸可以吃更多的美餐？”

“哦，是呀！鲸一到北极地区，就开始着手打造自己的菜园。”

“那我们就是鲸的菜园里的菜了！好主意，饲养你就是为了吃掉你。”

“So, are you telling me that because whales eat us, they create blooms of plankton, and that this leads to more of you krill being around? So the whale can have bigger dinners?”

“Oh yes! As soon as the whales arrive in the Arctic, they start tending their gardens.”

“And we are their gardens! Nice idea, that you are fed – just to be eaten.”

“看，这和菜园里的蔬菜没什么不同。是时候让我们开始用不同的方式来思考海洋生活了。这不仅仅是猎物和捕猎者的关系。这一切都是为了提高恢复能力和效率。”

“我才不关心大吃小、少吃多的问题。我只关心一个问题：活着的意义就是被吃掉。我觉得我不会像你那样感到兴奋。”

“Look, it is no different to vegetables in a garden. It is about time we started thinking differently about life in the sea. It is more than just prey and predators. It is all about more resilience and efficiency.”

“I don’t care if it is a matter of a big one eating many small ones, or a few eating the many. It all comes down to the same thing: live to be eaten. I don’t think I can get as excited as you about that.”

活着的意义就是被吃掉……

Live to be eaten ...

……捕杀鲸以获取鲸油……

... killing whales for their oil ...

“但这不一样。试想如果陪伴你们的只有几头鲸和几只磷虾……”

“这正是人们开始捕杀鲸以获取鲸油时发生过的事情。只剩下少量的鲸，进而导致磷虾和我们这些浮游生物也所剩不多……”

“But this is different. Imagine you start with just a few whales and a few of us…”

“That is exactly what happened when people started killing whales for their oil. There were only a few left, and as a result there were only small numbers of you krill left, and of us plankton… ”

“后来，人们意识到，他们已经把鲸逼到灭绝的边缘。一旦人们停止捕杀鲸，鲸的数量又回升了，现在再看看鲸……”

“通过向周围环境大量施肥来种植自己的食物，这其中就包括你和我。我不相信。”

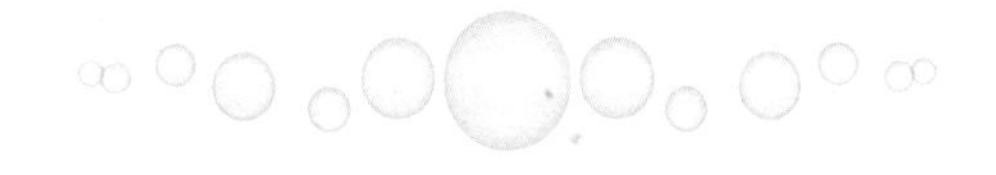

“And then, people realised that they have been pushing the whales to near extinction. As soon as they stopped killing them, the number of whales bounced back, and look at them now...”

“Wildly fertilising their surroundings – to grow their own food, including you and me. I am not convinced.”

……鲸的数量又回升了……

… the number of whales bounced back …

这些鲸真是高超的农夫啊！

These whales are mighty farmers!

“那你知道吗，鲸在海里潜下去又浮上来，在这个过程中，那些落在海底的零星铁质又被鲸从海底带上来了。当他们深潜时，他们的腹部会承受更多的压力，这迫使他们更想排便！”

“当他们出现在水面上时，他们会放松自己，选择在一个阳光充足的地方排出那些大便。然后我们就可以捕获能量了。这些鲸真是高超的农夫啊！”

“And did you know, those tiny bits of iron that drop to the ocean floor, are pulled back up by the whales diving down and coming up again. When they dive deep, there is more pressure on their tummies, making them want to poo!”

“And the moment they pop up to the surface, they relieve themselves, releasing it all exactly where there is plenty of sun. And then we capture that energy. These whales are mighty farmers!”

“是的，鲸是世界上最好的循环大师。他们的菜园是如此茂盛，然而人们现在才意识到鲸在海洋中的生产力比陆地上的任何作物都要多。”

“我们是鲸优秀的合作伙伴。我们一起维持循环，创造生命——即使鲸真的会把我们活活吃掉。”

……这仅仅是开始！……

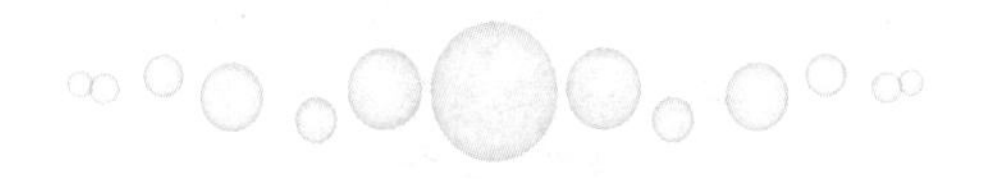

“Yes, whales are the best recyclers in the world. Their gardens are so lush, and yet people are only now realising how much whales produce in the ocean – so much more than any crop on land.”

“And we are their proud partners. Together we keep the cycle going, creating life – even if they do eat us alive.”

... AND IT HAS ONLY JUST BEGUN!...

……这仅仅是开始！……

... AND IT HAS ONLY JUST BEGUN! ...

Did You Know ?

你知道吗？

Krill is the most fished animal in the Southern Oceans. Their oily bodies are used in pharmaceuticals as a source of omega-3, and fed to livestock and aquaculture fish. It is also widely used in pet food.

磷虾是南大洋中被捕获最多的动物。由于能提供 Ω-3 脂肪酸，它们富含油脂的身体被用于制药，它们也被用作牲畜养殖和水产养殖的饲料。磷虾还被广泛用于宠物食品。

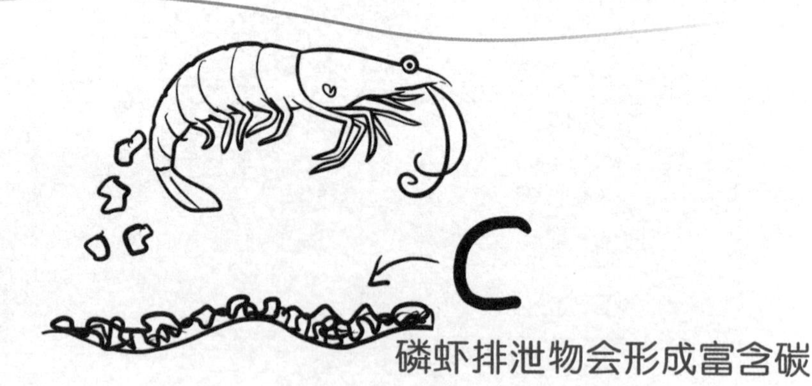

Krill excrement forms carbon-rich pellets that quickly sink to the ocean depths, where it may remain for many years. It locks carbon away from the atmosphere for long periods of time.

磷虾排泄物会形成富含碳的球状小颗粒，很快沉入海洋深处，可能会在那里停留多年。这些小颗粒能长时间把碳与大气环境隔离开。

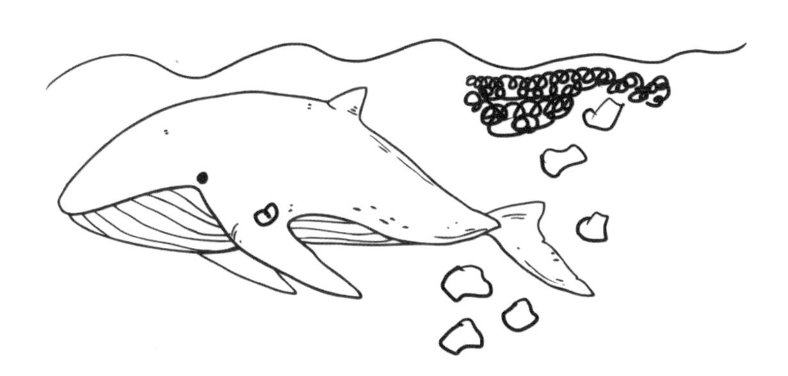

鲸的排泄物常发挥肥料的功能，可以增加浮游植物清除大气中二氧化碳的能力。微小的颗石藻是自然界消耗大气二氧化碳最多的生物之一。

Whale excrement acts as a fertiliser that increases the capacity of phytoplankton to remove atmospheric carbon dioxide. The tiny coccolithophores are one of nature's most prolific consumers of atmospheric carbon dioxide.

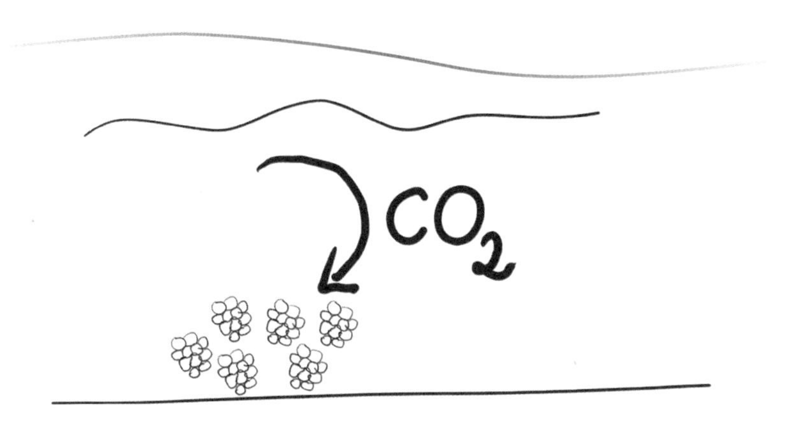

当浮游植物死亡时，残余物质会沉入海洋深处。据估算，浮游植物每年将 20 亿吨二氧化碳封存到海洋中，约占地球上所有封存碳的 90%。

When phytoplankton perishes, remains sink into the ocean depths. Phytoplankton sequesters an estimated 2 billion tons of CO_2 into the ocean each year, holding an estimated 90% of all sequestered carbon on Earth.

Whales form part of a positive feedback loop. As whale populations recover in the oceans, greater phytoplankton productivity will result in larger amounts of iron being recycled throughout the system.

鲸是正反馈循环的一个组成部分。随着海洋中鲸的数量回升，更强大的浮游植物生产力会使整个系统中更多的铁被循环利用。

Iron recycling by large baleen whales was reduced 10-fold by whale hunting between 1900 and 2008. Micro-zooplankton contributes most to biological iron recycling, followed by carnivorous zooplankton and krill.

1900 年至 2008 年间，因捕鲸活动，大型须鲸所带来的铁循环量减少了 90%。微型浮游动物对铁循环的贡献最大，其次是肉食性浮游动物和磷虾。

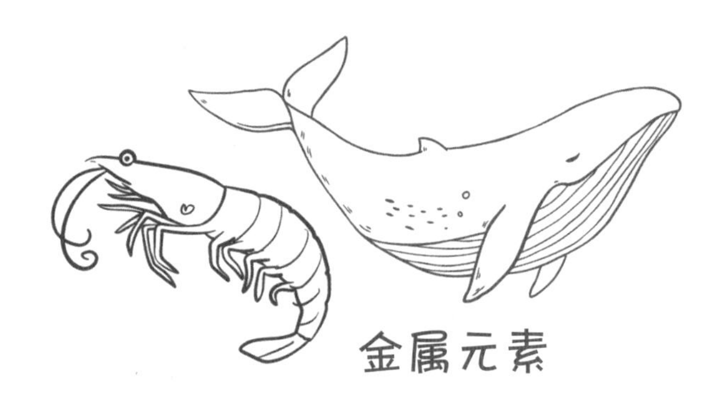

金属元素

Metal concentrations in krill tissue are between 20,000 and 4.8 million times higher than in seawater, while in whale faecal matter it is between 276,000 and 10 million times higher.

磷虾身体组织中的金属元素含量是相同质量海水的 2 万至 480 万倍，而鲸的粪便中的金属元素含量是相同质量海水的 27.6 万至 1 000 万倍。

Mn Zn Fe Cu P Cd C Co

The availability of micronutrients is a key factor affecting productivity of the food cycles. This includes iron, cadmium, manganese, cobalt, copper, zinc, phosphorus and carbon.

能否获取足够的微量营养元素是影响食物循环生产力的一个关键因素。这些微量营养元素包括铁、镉、锰、钴、铜、锌、磷和碳。

Think about It

想一想

Can poo really be that nutritious?

便便真的那么有营养吗?

Are these whale gardens real or imaginary?

这些鲸的菜园是真实存在的还是想象出来的?

Is life with more species more abundant? Or does this overpopulation lead to more hunger and poverty?

更多的物种会使生态系统更丰饶吗? 还是说，数量过多会导致更多的饥饿和贫困?

Does Nature celebrate the life of only one being, or of everyone?

大自然只赞美一个物种的生命，还是每个物种的生命?

Do It Yourself!

自己动手！

The increase in population in whales leads to an increase in population in krill and phytoplankton, and it even increases the capacity to sequester carbon dioxide from the atmosphere. This is the opposite of what we typically consider as the norm: over-population leads to poverty. Why human overpopulation of the Earth creates so many problems, while an "overpopulation" of whales generates more life and resilience? Discuss the conclusions you have come to, and ask each other what it is that people have to do to accommodate an increase in population.

鲸的数量增加导致磷虾和浮游植物的数量增加，甚至增强了这一生态系统封存大气中二氧化碳的能力。这与我们通常认为的正常情况恰好相反：人口过多导致贫困。为什么地球上人类过多会造成如此多的问题，而鲸的“过剩”反而会产生更多的生命和活力？讨论一下你得出的结论，并询问彼此人们需要做些什么来适应人口的增长。

学科知识

Academic Knowledge

生物学	像鲸这样的哺乳动物的新陈代谢，会释放数十亿生物所需的微量营养物质；在南极，有8种鲸吃磷虾；露脊鲸吃浮游动物；南极磷虾是已知的85种磷虾中最大的一种；磷虾长着坚硬的外骨骼和许多条腿；微生物占海洋生物量的98%，它们的总称是浮游生物。
化　学	铁元素是一种重要的营养物质；磷虾含有一种名为虾青素的消炎色素；磷虾的氨基酸有谷氨酸/谷氨酰胺、天冬氨酸/天冬酰胺、甘氨酸、丙氨酸、赖氨酸和亮氨酸；海洋浮游生物富含成岩物质；浮游植物含有高浓度的钙、磷、锰、铁、锌、铝、钡和铅。
物　理	水中的光和温度随海水深度的变化而变化，这决定了浮游植物的生长，或者有潜在的生理学相关性；“撇食”和胸足。
工程学	寒冷水域的生产效率为60亿吨南极磷虾提供了居住条件；磷虾成群结队地巡游，将此作为一种防御机制。
经济学	规模经济和范围经济；一头蓝鲸一天能吃掉4吨磷虾；雌性磷虾一次能产卵1万个，创造了高产的循环。
伦理学	食物链支撑着生命系统，而目前的渔业和农业减少了可用于强化生命系统的营养物质；将生命视为一个系统的能力与将生命视为单个物种的能力；一个物种的生命比整个生态系统的生命更重要吗？
历　史	提尔人马里纳斯在公元2世纪给南极洲命名；1820年俄罗斯探险队发现南极洲；挪威人罗阿尔德·阿蒙森1911年到达南极极点。
地　理	须鲸在夏季大约有4~6个月的时间在高纬度、多产的水域密集觅食；磷虾成群结队地聚集，密集且分布广泛，以至于在太空中都可以看到它们；南极洲曾被称为未知的南方大陆。
数　学	虽然每只磷虾只有2克重，但所有磷虾的总质量估计有20亿吨。
生活方式	我们把生物简化为猎物和捕食者，但生命不仅仅是吃和被吃；磷虾有很强的适应力。
社会学	只关注自己的利益和生存机会，与有机会看到对所有生物都有益的东西（甚至是不属于同一物种的其他个体）。
心理学	先入为主的想法：如果你认为某件事是坏的，即使事实证明它是好的，你也会把它当作坏的；是什么让你感到骄傲。
系统论	一个生态系统中的更多物种是怎样产生更多的营养物质来支持更丰富的生命；每个个体在大生态系统中的作用；南极磷虾和蓝鲸在南大洋生态系统中扮演着关键角色；浮游植物贡献了全球一半的初级生产力，是水生环境中食物网的基础，在全球碳循环的反馈循环中起着关键作用；自20世纪70年代以来，南极磷虾的储量可能下降了80%。

情感智慧

Emotional Intelligence

磷　虾

磷虾自发地表达了自己的兴奋，让每个人都能听到。她对鲸所起到的作用理解得非常透彻，尤其能理解鲸的排泄物如何供养更多的生命。她了解这个错综复杂的生命网络的细节，并毫无保留地分享自己的知识。她选用简单的隐喻来让浮游生物更好地理解，并督促他重新思考审视海洋生命的方式。她很有耐心，一步一步地向对方解释。当浮游生物还没有掌握营养物质循环的本质时，她会花更多的时间来解释，并追溯历史。她对事实的陈述使浮游生物能够更好地理解他们是如何在生命系统中扮演重要角色。

浮游生物

浮游生物明确表示自己对磷虾的兴奋并不认同。他毫不客气地反对磷虾之前提出的逻辑，还点明了小物种要被大物种吃掉的事实。他用一连串的事实展示了他自己的逻辑，并得出结论：为了其他物种的生存，有些物种必须死。他对自己的分析深信不疑，不容易被磷虾所左右。然而，当磷虾开始解释在原本濒临灭绝的鲸的数量恢复后，他们所处的生态系统是如何恢复生机的，浮游生物便发现了更多由太阳提供的无限能量所驱动的水中营养物质的良性循环。

艺术

The Arts

如果只运用逻辑和科学，很难掌握复杂的生命循环。重要营养物质和能量的流动，例如鲸的排泄物和太阳作用，形成了生命循环的基础。让我们用艺术手法来表现这两个因素的结合是如何创造丰富生命的。这种结合激发了地球上一些极小但又极为丰富的动物生命形式，比如磷虾和浮游动物。

思维拓展

Systems: Making the Connections

地球生命捕获了两种来源的能量：物质循环中的养分和来自太阳的能量。对于粪便存在着一种简单的看法，那就是将其归类为废物，并认为它对健康有害。然而，这种观点太过狭隘。这类阐释会导致资源浪费，进而引发饥饿和贫困。我们需要更好地理解如何通过改善营养、物质和能量的循环，来扩大生命的密度和丰富度。一种生命体产生的废物可能会对另一种生命体大有裨益。鲸的排泄物可以供养其他生命体，就像大树不直接回收自己的树叶，而是把它们留给其他生物。这不是一个单纯的因果关系，而是一套复杂的营养循环体系。生物网中所有成员间的持续互动，创造着一种适应力，让生命得以繁衍。我们知道的是，当食物网中物种构成不变而营养循环增强时，生物的密度就会增加。这就是为什么更多的人类意味着更多的贫困，因为我们还没有形成闭环。大自然使我们重新思考如何消除世界范围内的饥饿问题。据统计，仍有10多亿人遭受营养不良和饥饿的困扰。尽管粮食产量有所增加，但我们未能成功地循环利用所有可利用的养分，为更多的人生产更多的粮食。海洋中可供鲸漫游的空间是陆地上可利用空间的数倍。海洋和陆地之间最大的区别在于，海洋是三维的，而陆地是二维的，因此海洋能提供更广阔的发展空间。另外，密度也有差别。水的密度是空气的几百倍，因此在同样的空间里，水比空气可以容纳更多的营养物质。通过遗传学和化学技术的应用，在海洋中可实现的生产力将是陆地上的数倍。

动手能力

Capacity to Implement

与海洋相比，我们在陆地上的效率能有多高？为了转变我们对未来的认知，确保向更可持续的社会过渡，我们需要知道最大的潜力在哪里。如果我们考量三维空间的利用率和营养物质的密度，那么陆地与海洋相比将相形见绌。为了今后能将这些设想付诸实践，让我们从计算开始。查找资料并试着计算，海洋的生产力是陆地的多少倍？

故事灵感来自

This Fable Is Inspired by

埃莉诺·贝尔
Eleanor Bell

埃莉诺·贝尔在苏格兰圣安德鲁大学获得环境生物学学位，由此开始了科研生涯。后来，她又在英国诺丁汉大学获得博士学位。求学期间，她花了相当长的时间在南极洲的澳大利亚戴维斯科考站，研究盐湖中的微生物食物网。在麦吉尔大学（加拿大）的博士后工作期间，她研究海马。贝尔博士随后转到波茨坦大学（德国），在那里，她花了6年时间研究极端环境中的微生物生态学。2012年，她开始在澳大利亚海洋哺乳动物中心工作，该中心是澳大利亚政府南极分部的一个组成部门。

图书在版编目(CIP)数据

冈特生态童书.第七辑:全36册:汉英对照 /
(比)冈特·鲍利著;(哥伦)凯瑟琳娜·巴赫绘;
何家振等译.—上海:上海远东出版社,2020
ISBN 978-7-5476-1671-0

Ⅰ.①冈… Ⅱ.①冈… ②凯… ③何… Ⅲ.①生态
环境-环境保护-儿童读物—汉英 Ⅳ.①X171.1-49

中国版本图书馆CIP数据核字(2020)第236911号

策　　划 张　蓉
责任编辑 祁东城
封面设计 魏　来 李　廉

冈特生态童书
鲸的菜园
[比]冈特·鲍利　著
[哥伦]凯瑟琳娜·巴赫　绘
田　烁　译